1,000,000 Books

are available to read at

www.ForgottenBooks.com

Read online
Download PDF
Purchase in print

ISBN 978-1-333-62171-1
PIBN 10527491

This book is a reproduction of an important historical work. Forgotten Books uses
state-of-the-art technology to digitally reconstruct the work, preserving the original format
whilst repairing imperfections present in the aged copy. In rare cases, an imperfection in
the original, such as a blemish or missing page, may be replicated in our edition. We do,
however, repair the vast majority of imperfections successfully; any imperfections that
remain are intentionally left to preserve the state of such historical works.

Forgotten Books is a registered trademark of FB &c Ltd.
Copyright © 2018 FB &c Ltd.
FB &c Ltd, Dalton House, 60 Windsor Avenue, London, SW19 2RR.
Company number 08720141. Registered in England and Wales.

For support please visit www.forgottenbooks.com

1 MONTH OF
FREE
READING

at

www.ForgottenBooks.com

By purchasing this book you are eligible for one month membership to ForgottenBooks.com, giving you unlimited access to our entire collection of over 1,000,000 titles via our web site and mobile apps.

To claim your free month visit: www.forgottenbooks.com/free527491

* Offer is valid for 45 days from date of purchase. Terms and conditions apply.

English
Français
Deutsche
Italiano
Español
Português

www.forgottenbooks.com

Mythology Photography **Fiction**
Fishing Christianity **Art** Cooking
Essays Buddhism Freemasonry
Medicine **Biology** Music **Ancient
Egypt** Evolution Carpentry Physics
Dance Geology **Mathematics** Fitness
Shakespeare **Folklore** Yoga Marketing
Confidence Immortality Biographies
Poetry **Psychology** Witchcraft
Electronics Chemistry History **Law**
Accounting **Philosophy** Anthropology
Alchemy Drama Quantum Mechanics
Atheism Sexual Health **Ancient History**
Entrepreneurship Languages Sport
Paleontology Needlework Islam
Metaphysics Investment Archaeology
Parenting Statistics Criminology
Motivational

UNIV. OF
CALIFORNIA

Uniformity in Invertase Action

By
DAVID INGERSOLL HITCHCOCK

DISSERTATION

SUBMITTED IN PARTIAL FULFILLMENT OF THE REQUIREMENTS FOR THE
DEGREE OF DOCTOR OF PHILOSOPHY IN THE FACULTY
OF PURE SCIENCE, COLUMBIA UNIVERSITY

New York City
1922

UNIV. OF
CALIFORNIA

Y FATHER AND MOTHER

UNIV. OF
CALIFORNIA

ACKNOWLEDGMENT

The author desires to express his sincere gratitude to Professor John M. Nelson for his kindly direction of this work. He wishes also to thank Dr. Warren C. Vosburgh for much friendly advice.

To the Harriman Research Laboratory, New York City, thanks are due for financial assistance which made this work possible.

D. I. H.

LABORATORY OF ORGANIC CHEMISTRY,
COLUMBIA UNIVERSITY
AUGUST, 1921

ABSTRACT OF DISSERTATION.

1. What was attempted?
2. How far were the attempts successful?
3. What contribution actually new to the science of Chemistry has been made?

1. (a) An attempt was made to determine whether or not different preparations of the enzyme invertase would cause the hydrolysis of cane to proceed at correspondingly identical rates in parallel experiments.

(b) A further attempt was made to find a general expression for the normal course of the hydrolysis of sucrose by invertase under varied conditions of enzyme concentration, temperature, and hydrogen· ion concentration.

(c) Attempts were also made to determine whether the action of different invertase preparations could be modified so as to make them all act alike.

2. (a) It was found that not all invertase preparations act quantitatively alike throughout the hydrolysis, a few preparations allowing the reaction to become abnormally slow after the first twenty per cent of the inversion.

(b) For the normal invertase preparations it was found that the hydrolysis-time curves coincided exactly if the proper amounts of invertase were used. An empirical equation was obtained and so transformed as to become generally applicable to hydrolyses in which different amounts of invertase were used. It was shown that within certain limits this equation represented the course of the hydrolysis not only for experiments in which the invertase concentration was varied, but also for experiments in which the temperature and hydrogen ion concentration were varied. It was also shown that this equation could be used as a criterion of normal invertase action.

(c) It was found that the abnormal course of the reaction due to one invertase preparation could be obviated by the addition of boiled normal invertase or of sodium chloride, while the action of another abnormal invertase preparation was not affected by these additions. It was found that a normal invertase preparation could not be rendered abnormal by further dialysis or by partial inactivation by heat or by ultraviolet light.

3. The new contributions made in the present work are as follows:

(a) Not all invertase preparations act identically in hydrolyzing sucrose, a few being abnormal in allowing the reaction to slow up after the first twenty per cent more than is the case with the normal majority of invertase preparations.

(b) For the hydrolysis of sucrose by normal invertase preparations, the time of reaction has for the first time been exactly expressed as a definite function of the per cent hydrolyzed. It has been shown that this function is of the same form for changes, within certain limits, in the invertase concentration, temperature, and hydrogen ion concentration. The equation expressing this function yields a constant which is the most satisfactory measure so far proposed of the true activity of the invertase.

(c) It has been found that the abnormality of one abnormal invertase preparation is removed by the presence of boiled normal invertase or sodium chloride, while that of another is not, and also that the normal invertase cannot be made abnormal either by dialysis or by the action of

CALIFORNIA

UNIFORMITY IN INVERTASE ACTION.

I. Introduction.

The study of the nature of enzymes has developed chiefly along two distinct lines. So far all attempts to isolate an enzyme as a pure substance of definite chemical composition have been without satisfactory results. Accordingly considerable work has been done to gain an insight into the nature of the enzyme itself by studying the velocity of the reaction which is catalyzed by the enzyme. The object in all these researches on the chemical kinetics of enzyme action has been to find some general law governing the rate of the reaction, and by means of this to make deductions concerning the mechanism of the action and the nature of the enzyme.

In all such work the tacit assumption has been made that any two preparations of the same enzyme, if the same amount of active enzyme is present, and other conditions are the same, will catalyze the reaction at identical rates at corresponding points throughout. In other words, if the course of the reaction were plotted, the two curves should be superimposable. While this assumption has not been definitely stated in the past, it is a necessary implication of the attempt to find a general law for the rate of the catalyzed reaction. Accordingly it seemed desirable, before going further in the attempt to obtain a general law for the hydrolysis of cane sugar by the catalytic action of invertase, to find out by direct experiment whether several different enzyme preparations would really give the same quantitative course to the reaction.

II. Differences in Invertase Action.

In order to make this comparison, it was necessary to have all the conditions alike except for the use of different invertase preparations. The conditions adopted were a temperature of 25°, an initial sucrose concentration of 10 g. per 100 cc., and a hydrogen-ion concentration of $10^{-4.4}$ to $10^{-4.5}$ secured by 0.01 M buffer solution of acetic acid and sodium acetate. The amounts of invertase were so adjusted in each experiment that the reaction started off at the same rate. This was accomplished by a few preliminary experiments, making use of the fact noted by Nelson and Vosburgh,[1] that the velocity of inversion is directly proportional to the concentration of the enzyme. The extent of inversion was determined by the polariscopic method. That this method is justifiable has been established by Vosburgh[2]

[1] Nelson and Vosburgh, *J. Am. Chem. Soc.*, **39**, 790 (1917).
[2] Vosburgh, *ibid.*, **43**, 219 (1921).

Table I contains the results of 10 experiments with 4 different invertase preparations. These preparations were all obtained from yeast by modifications of the method of Nelson and Born. No differences in the method of preparation are known which might account for the abnormality of No. 3. Further description of the method of preparation will be found under the heading "Experimental Details."

TABLE I.
DIFFERENCES IN INVERTASE ACTION.

Temp., 25°. Conc. of sucrose, 10 g. per 100 cc. Conc. H$^+$, 10$^{-4.4}$ to 10$^{-4.5}$ moles per liter.

A. Normal Invertase.

Expt. Invertase Preparation No.	B1. 8.	B2. 8.	B5. 2.	B6. 2.	B9. 1.	B10. 1.	Mean.
Cc. of Invertase per 100 cc.	1.530.	1.530.	7.141.	7.141.	6.080.	6.080.	
Initial Rotation, degrees.	13.05.	13.04.	13.17.a	13.18.a	13.05.	13.04	

Time t. Min.	Change in rotation, degrees.							Inverted %.
5	0.54	0.54	0.53	0.54	0.52	0.51	0.53	3.15
10	1.07	1.07	1.07	1.07	1.06	1.06	1.07	6.35
15	1.58	1.59	1.58	1.59	1.58	1.57	1.58	9.38
22	2.33	2.34	2.32	2.31	2.31	2.31	2.32	13.77
30	3.12	3.14	3.12	3.12	3.12	3.12	3.12	18.52
60	5.97	5.98	5.97	5.96	5.98	5.97	5.97	35.43
90	8.47	8.48	8.46	8.44	8.50	8.49	8.47	50.27
120	10.57	10.59	10.55	10.54	10.58	10.59	10.57	62.73
180	13.52	13.53	13.53	13.50	13.55	13.53	13.53	80.30
300	15.93	15.93	15.91	15.93	15.93	15.91	15.92	94.48
2 to 4 days	16.87	16.86	16.85	16.87	16.85	16.85	16.86	.....

B. Abnormal Invertase.

Expt. Invertase Preparation No.	B12. 3.	B13. 3.	B14. 3.	B15. 3.	Mean.
Cc. of Invertase per 100 cc.	1.905.	1.905.	1.905.	1.905.	
Initial Rotation, degrees.	13.05.	13.05.	13.05.	13.05.	

Time. t. Min.	Change in rotation, degrees.					Inverted %.
15	0.55	0.54	0.54	0.54	0.54	3.20
10	1.07	1.08	1.07	1.07	1.07	6.35
15	1.59	1.59	1.59	1.57	1.59	9.44
22	2.30	2.31	2.29	2.30	2.30	13.65
30	3.13	3.11	3.10	3.08	3.11	18.46
60	5.92	5.87	5.88	5.88	5.89	34.96
90	8.33	8.30	8.32	8.31	8.32	49.38
120	10.39	10.34	10.36	10.35	10.36	61.48
180	13.27	13.26	13.30	13.28	13.28	78.81
300	15.77	15.73	15.78	15.81	15.77	93.58
2 to 4 days	16.85	16.85	16.85	16.85	16.85	....

a The high values of these figures are due to the fact that invertase No. 2 itself had an unusually high rotation.

It will be noticed that the results indicate that in 6 experiments with 3 different invertase preparations the invertase is acting alike throughout the course of the reaction. The results of the last 4 experiments, on the other hand, while agreeing among themselves, do not agree well with the other results beyond about the first 20% of the inversion. This indicates that invertase preparation No. 3 is in some way different from the other preparations, since the latter part of the reaction is noticeably slower. These results therefore show that not all preparations of invertase have, under a given set of conditions, the same degree of activity. It was thought at first that this retardation might be due to spontaneous destruction of the enzyme. That this was not the case is shown by the results of experiment B15, for in this case the invertase was kept in the thermostat for 5 hours before the start of the reaction. It has been shown by O'Sullivan and Tompson[3] that invertase suffers spontaneous destruction less rapidly in the presence of sucrose than in its absence. Since the results of Expt. B15 indicate no loss in activity of the invertase in 5 hours without the presence of sucrose, it is evident that the difference in the action of invertase preparation No. 3 cannot be due to spontaneous destruction of the enzyme. Table I proves, therefore, that while there seems to be a normal course for invertase action, there are also exceptions or abnormal invertase preparations.

III. Discussion of Equations for Invertase Action.

If a general equation for normal invertase action were available, it would be comparatively easy to ascertain from experimental data whether any given preparation were normal or abnormal. Nelson and Vosburgh[1] and others have shown that the rate of hydrolysis of cane sugar by invertase is not proportional to the concentration of the substrate, or, in other words, the reaction does not obey the unimolecular law,

$$k = \frac{1}{t} \log \frac{a}{a-x} \tag{1}$$

where k is the velocity coefficient, t time in minutes, a the initial cane sugar concentration and x the amount hydrolyzed.

Henri[4] proposed an empirical equation containing only one constant.

$$k = \frac{1}{t} \log \frac{a+x}{a-x}. \tag{2}$$

When this equation was applied to the results with normal invertase given in Table IA, decreasing values for the constant were obtained.

Equations of the form

$$t = k_1 \log \frac{a}{a-x} + k_2 x. \tag{3}$$

[3] O'Sullivan and Tompson, *J. Chem. Soc.*, **57,** 834 (1890).
[4] Henri, "Lois générales de l'action des diastases," Hermann, Paris, **1903**, p. 59.

were deduced by Henri[5] and Bodenstein,[6] by Barendrecht,[7] Michaelis and Menten,[8] and Van Slyke and Cullen.[9] On applying the method of least squares to the present results (Table IA) to determine the constants for an equation of this type, it was found that such an equation would not hold satisfactorily for the whole course of the reaction. However, by using the values obtained for the first half of the hydrolysis only, the following equation, obtained by least squares, was found to hold for the first 50% of the hydrolysis.

$$t = 135.2 \log \frac{100}{100-p} + 0.9724\ p. \qquad (4)$$

Here t is the number of minutes required for p per cent. of the sucrose to be inverted. When p is expressed in the notation of Equation 1 it is equal to 100 x/a. The accuracy with which the equation fits the results may be seen from the following figures.

$p.$	t (calc.).	t(obs.).	$p.$	t(calc.).	t(obs.).
3.15	4.94	5	18.52	30.0	30
6.35	10.0	10	35.43	60.1	60
9.37	14.9	15	50.27	89.9	90
13.77	22.1	22			

By considering the hydrolysis of sucrose as a reversible reaction, unimolecular in one direction and bimolecular in the other, Visser[10] deduced an equation which, in the symbols previously used, takes the form

$$\frac{dx}{dt} = k_1(a-x) - k_2 x^2. \qquad (5)$$

On applying this to his experiments he obtained increasing values for a quantity which should theoretically have been constant. In order to correct for this increase he introduced a factor I to which for some reason he attributed a chemical significance, and called it "the intensity of the enzyme," giving the equation:

$$\frac{dx}{dt} = [k_1(a-x) - k_2 x^2]I. \qquad (6)$$

The values of I he obtained from the increasing values of the constant obtained from Equation 5, and he found they could be expressed by the empirical formula

$$I = \frac{k_3}{7a^2 - 4ax + x^2}. \qquad (7)$$

By substituting this expression in Equation 6 and integrating he obtained

[5] Op. cit., p. 79.

[6] Bodenstein, ibid., p. 92.

[7] Barendrecht, Z. physik. Chem., 49, 456 (1904).

[8] Michaelis and Menten, Biochem. Z., 49, 333 (1913).

[9] Van Slyke and Cullen, J. Biol. Chem., 19, 141 (1914).

[10] Visser, Z. physik. Chem., 52, 257 (1905).

a complicated equation which gave constants satisfactory to about $\pm 10\%$. Since Hudson[11] has shown Visser's theory of reversibility to be unsound, no attempt was made to apply this equation to the present results.

Visser also proposed a simpler equation for invertase action, obtained by neglecting the reverse reaction, but putting the same factor I into the unimolecular equation.

$$\frac{dx}{dt} = k(a-x)I = k(a-x)\frac{k_3}{7a^2 - 4ax + x^2}. \tag{8}$$

This on integration becomes

$$2k_1 k_3 t = 8a^2 \ln\frac{a}{a-x} + x(6a-x) \tag{9}$$

Visser's application of this equation to his own experiments gave "constants" which showed an extreme variation of 30% (0.00108 to 0.00139) and an application of the same equation to the experiments of Table IA likewise gave increasing values for the constant.

An equation of similar form,

$$t = k_1 \log\frac{a}{a-x} + k_2 x + k_3 x^2, \tag{10}$$

was obtained from the data of Table IA by the method of least squares, but did not fit the experimental data well enough to be of any use in the present work.

In order to get satisfactory agreement with the experimental data, it was found necessary to use an equation containing four constants. The following equation was obtained by applying the method of least squares to the mean results of the 6 experiments in Table IA.

$$t = 222.9 \log\frac{100}{100-p} + 0.5890p - 0.001975p^2 - 0.00002034p^3. \tag{11}$$

The applicability of this equation may be seen from the following figures.

TABLE II.

APPLICATION OF EMPIRICAL EQUATION.

Inversion. p %.	Time. t(calc.). Min.	Time. t(obs.). Min.	$n \times 10^{5}$.[a]
3.15	4.96	5	443
6.35	10.1	10	449
9.38	14.9	15	445
13.77	22.1	22	449
18.52	30.1	30	447
35.43	60.0	60	446
50.27	89.8	90	445
62.73	119.6	120	445
80.30	180.6	180	448
94.48	299.9	300	447
		Mean,	446
		A. d.,	0.36%.

[a] Meaning of n explained below.

[11] Hudson, *J. Am. Chem. Soc.*, **36**, 1571 (1914).

IV. An Equation for Experiments with Different Amounts of Invertase.

Nelson and Vosburgh[1] showed that, in their experiments in which the initial sucrose concentration was constant, the time required for a given percentage of sucrose to be inverted was inversely proportional to the amount of invertase used. In other words, letting t represent the time for 80% inversion, and y the invertase concentration, in experiments in which only the invertase concentration was varied they found the product ty to be constant. In another series of experiments with a different invertase preparation they called t the time for 40% inversion, and here again ty was constant. They did not, however, compare the times for different degrees of inversion in any one experiment, say t for 40% and t for 80% inversion, because they did not know the law governing the relationship between the time and the percentage of inversion, or, mathematically, the form of the function, $t = f(p)$.

It is obvious that one may plot the values for the amounts hydrolyzed in various times against the times, obtaining curves for the hydrolyses which are graphical representations of the function, $t = f(p)$. This was done by Michaelis and Davidsohn[12] in such a way as to compare the form of the function in three experiments with invertase concentrations in the ratio 2:1:0.4. They plotted the product of the enzyme concentration and the time, ty, against the change in rotation, which is proportional to the percentage inverted, p. They claimed that the points all fell on a smooth curve, and that therefore the form of the hydrolysis curve was independent of the amount of enzyme. However, only 3 experiments were given of which one was represented by only 2 points. Moreover, their whole curve did not appear to extend much beyond the first half of the inversion, and in addition several of their points did not fall on the curve, even on the small scale used in their printed article. Because of these deficiencies, and because the shape of the curve is a fundamental point in the present investigation, it seemed best to amplify their data by the use of the more extensive experiments of Nelson and Vosburgh.

Accordingly the results of the latter were plotted in a similar way on a large scale. The curves were brought together at one point by using a different time scale for each experiment. When the remainder of each curve was plotted on this new scale, it was found that the curves for experiments with the same initial sucrose concentration did superimpose, falling on a single smooth curve. Thus the conclusion drawn by Michaelis and Davidsohn was more firmly established by the results of Expts. 6, 7, 8, 9, 10, 22 and 23 of Nelson and Vosburgh, each experiment including at least 6 samples and extending over 95% or more of the inversion. This means that the function, $t = f(p)$, representing a single experiment, can be generalized as $nt = F(p)$ for experiments with varying amounts of invertase.

[12] Michaelis and Davidsohn, *Biochem. Z.*, **35**, 386 (1911).

Here n is a constant in any one experiment, but varies in different experiments, being proportional to the amount of effective invertase. Moreover the form of the function $nt = F(p)$ is, within the limits of these experiments, independent of the amount of invertase or of the rate of the hydrolysis.

A more exact verification of this relationship was obtained by the use of Equation 11, which gives a definite form to the function, $t = f(p)$, for one particular invertase concentration. In order to make this equation generally applicable to experiments with other invertase concentrations, the coefficient of the logarithmic term was placed equal to $1/n$ and factored out, giving

$$t = \frac{1}{n} \left[\log \frac{100}{100-p} + 0.002642p - 0.000008860p^2 - 0.0000001034p^3 \right]. \quad (12)$$

If it is generally true that the times for any given percentage of inversion are inversely proportional to the amounts of invertase used, then Equation 12 gives a definite form to the function, $t = f(p)$, for any invertase concentration. Whether or not this is the case can be tested by substituting in Equation 12 the experimental values for p and t, and calculating the values of n. If the latter are constant, the equation applies and the general law holds, and the values of n should be directly proportional to the amounts of active invertase present.

To recapitulate, we have, if this is true, first an empirical relation between time and percentage inverted which holds for experiments in which different amounts of invertase are used; and, second, we have in the value of n a relative measure of the amount of the effective invertase.

In order to decide whether or not Equation 12 applies to a given experiment, it is necessary to decide whether or not the values of n are constant. The values of n for the experiments of Table IA, from which the equation was derived, are given in the last column of Table II. The average deviation from the mean, 0.36%, is an indication of the extent to which the equation fits these original experiments. To determine about what magnitude of deviation from the mean might be due to experimental error in applying the equation to other experiments, the following calculations were made.

From the agreement of duplicate experiments in Table I and subsequent experiments, the average error in determining any change in rotation was estimated as 0.02°. To determine what error in n could be caused by such an error, the value 0.02° was added to all the changes in rotation of Table IA, and the values of n recalculated, with the results shown in Table III.

Evidently the form of the relationship is such that errors are magnified in the values of n calculated from the data on the early part of the hydrolysis. Since in all the experiments of the present work except those of Table I the first sample was taken at about 10% inversion, while the

TABLE III.

EFFECT OF ASSUMED ERROR OF 0.02°.

p + error.	t.	10^5 (n + error).	10^5 (n true values).	Error in $10^5 n$.	Dev. from true mean.
3.26	5	458	443	15	12
6.47	10	457	449	8	11
9.50	15	450	445	5	4
13.89	22	453	449	4	7
18.64	30	450	447	3	4
35.55	60	448	446	2	2
50.39	90	446	445	1	0
62.85	120	446	445	1	0
80.42	180	450	448	2	4
94.60	300	450	447	3	4

Mean, 4.4 = 0.99%.

Mean, 4.8 = 1.08%.

other 7 samples were distributed about as before, it was decided that a fairer measure of the average error in n would be given by the mean of the last 8 of the above values. This gives an average deviation of 0.59% from the individual values of n, or of 0.70% from the mean value of n as the deviation caused by an error of 0.02° in the value of each change in rotation. Hence it may fairly be decided that any experiment giving an average deviation from the mean of 0.7% or less is fitted by the equation, and its curve has the same shape or the function, $nt = F(p)$, has the same form as in the case of the original experiments of Table IA.

Since the results of Nelson and Vosburgh were available, including experiments in which the concentration of invertase was varied, it was thought that these results might well be used as a test of the general applicability of Equation 12. Accordingly Table IV was prepared by using those of their experiments in which the initial sucrose concentration was 10 g. per 100 cc.

In the last three of these experiments the average deviation from the mean of the values of n is below the value 0.7%. As has been already pointed out, this deviation might be caused by experimental error, and accordingly the equation fits these three experiments satisfactorily. In Expts. 6, 8, and 9 the first sample was taken before the inversion was 10% complete. Now it has been already pointed out that in this part of the inversion a small experimental error may cause a large error in the value of n. Accordingly for these experiments the mean of the remaining values of n was calculated, omitting the first, and the average deviations were found to be 0.64%, 0.40%, and 0.10% for Expts. 6, 8 and 9, respectively. Therefore the equation really does fit 6 of the 7 experiments in Table IV, and it may be concluded with more certainty than before that the shape of the hydrolysis curve or the form of the function, $nt = F(p)$, is independ-

TABLE IV.

EFFECT OF VARYING AMOUNTS OF INVERTASE. (EXPERIMENTS OF NELSON AND VOS-
BURGH.)

Initial sucrose concentration, 10 g. per 100 cc. Hydrogen-ion concentration, 3.2×10^{-5} to 2.1×10^{-5} moles per liter. Temperature, $37°$. Invertase preparation A used in Expts. 6 to 10, Preparation B in Expts. 22 and 23.

Expt.	Inv. per 100. Cc.	Time. t. Min.	Amt. inverted, p,%.	$n \times 10^5$.	Expt.	Inv. per 100. Cc.	Time t. Min.	Amt. inverted, p, %.	$n \times 10^5$.
6	6	14	8.32	422	7	5	22	11.08	360
		30	17.92	433			40	20.15	366
		70	39.85	437			90	43.21	372
		120	61.90	437			138	60.98	372
		185	80.59	439			215	80.30	375
		320	94.97	431			373	95.04	370
			Mean,	433				Mean,	369
			Av. dev., 1.04%.					Av. dev., 1.14%.	
8	4	20	8.31	295	9	3	33	9.57	206
		45	18.69	301			70	20.49	213
		105	41.17	302			150	41.50	213
		175	62.11	301			250	62.44	212
		265	80.09	303			376	80.09	213
		450	94.85	305			660	95.28	213
			Mean,	301				Mean,	212
			Av. dev., 0.73%.					Av. dev., 0.80%.	
10	2	50	10.01	143	22	1	30	10.91	259
		100	20.14	146			60	21.50	261
		221	42.09	147			122	41.21	260
		315	56.10	146			200	61.49	260
		345	60.04	146			305	79.65	260
		570	81.60	147			600	96.08	$(248)^a$
		1365	98.06	$(131)^a$				Mean,	260
			Mean,	146				Av. dev., 0.17%.	
			Av. dev., 0.57%.						
23	2	14	10.34	526					
		28	20.49	533					
		60	41.81	538					
		100	62.62	533					
		156	81.70	537					
		280	95.70	$(517)^a$					
			Mean,	533					
			Av. dev., 0.60%.						

^{a} These values are for points beyond the limit of p, 95%, for which the equation was derived, and hence were not used in taking the mean.

ent of the invertase concentration, and that Equation 12 gives a definite form to this function, F (p).

In view of the fact that the extreme variation in the invertase concentration in these experiments was from 6 cc. to 2 cc. or in the ratio 3:1, while the range covered by the rather unsatisfactory experiments of Michaelis and Davidsohn was 5:1, it seemed best to try the effect of a wider variation in invertase concentration on the shape of the curve. The highest concentration used was selected so as to make the hydrolysis as rapid as possible without causing error in the timing of samples, and the lowest concentration was such that the first and last samples could just conveniently be taken on the same day. In view of the difficulty encountered in a previous investigation[13] in obtaining reproducible results with very dilute invertase solutions, it seemed unwise to attempt to study slower reactions than this. The results of the experiments with the extreme invertase concentrations used, in the ratio 12:1, are given in Table V.

TABLE V.
EXTREME CHANGES IN INVERTASE CONCENTRATION.

Expts. B60 and B61.					Expt. B62.			
6 cc. of Invertase 8 per 100 cc.					0.5 cc. of Invertase 8 per 100 cc.			
Time, t. Min.	Rotation, B60. Degrees.	B61.	Amt. inverted, p, %.	$n \times 10^4$.	Time. t. Min.	Rotation, degrees.	Amt. inverted, p, %.	$n \times 10^5$.
0	13.09	13.09	.....	...	0	13.04	....	....
5	11.11	11.11	11.75	168	60	11.11	11.45	136
10	9.24	9.25	22.85	167	120	9.28	22.31	136
15	7.50	7.51	33.18	166	180	7.58	32.40	135
21	5.57	5.58	44.63	166	252	5.68	43.68	135
28	3.59	3.59	56.38	165	336	3.76	55.07	134
37	1.49	1.49	68.84	166	444	1.70	67.30	133
52	−0.93	−0.94	83.26	168	624	−0.70	81.54	134
70	−2.43	−2.43	92.11	170	840	−2.26	90.80	134
1–7 days	−3.76	−3.76	Mean,	167	11 days	−3.81	Mean,	135
			Av. dev.,	0.75%			Av. dev.,	0.65%

The values of n are sufficiently constant so that the equation may be said to hold for these concentrations.

If the time for any given percentage of inversion is inversely proportional to the concentration of invertase, the value of n divided by the number of cubic centimeters of invertase used per 100 cc. of solution should be a constant for any given invertase preparation. For Expts. B1 and B2 (Table IA) this value is 0.00292; for Vosburgh and Nelson's Expt. 1B (Table VI), it is 0.00290; for B60 and B61, 0.00278; and for B62, 0.00270. These experiments were all made with Invertase 8. The difference between the former two and the latter two values is due to slow deterioration of the invertase, even when kept in the ice-box, for a period of 8 months had elapsed between the two sets of experiments. The smaller difference between the latter two values can hardly be so explained, as the experiments were run on successive days, but must be taken to mean that for this range of concentrations the effective activity of the invertase is not strictly proportional to the actual concentration used. However, since the equation applies equally well in both cases, it may be stated as a fact

[13] Nelson and Hitchcock, "The Activity of Adsorbed Invertase," J. Am. Chem. Soc., 43, 1956 (1921).

that over this range of invertase concentrations (12:1) the form of the function, $nt = F(p)$, is the same and is expressed by Equation 12, while the value of n represents accurately the true activity of the invertase even better than its relative concentration.

V. Effect of Temperature.

Since the experiments of Nelson and Vosburgh were carried out at 37° while the present experiments were run at 25°, it seemed that the effect of temperature differences in any two experiments might be constant for all stages of the reaction. Inasmuch as some experiments on the course of the hydrolysis at various temperatures had recently been made in this

TABLE VI.

EFFECT OF TEMPERATURE. (EXPERIMENTS OF VOSBURGH AND NELSON.)

Initial sucrose concentration, 10 g. per 100 cc. Hydrogen-ion concentration, 4.4×10^{-5} to 4.0×10^{-5} moles per liter. Invertase preparation No. 8, 1 cc. per 100 cc.

Expt.	Temp. °C	Time, t. Min.	Amt. inverted, p, %.	$n \times 10^5$	Expt.	Temp. °C	Time, t. Min.	Amt. inverted, p, %.	$n \times 10^5$
11B	15	21	5.22	175	15B	20	16	5.04	222
		63	14.54	166			38	11.87	223
		110	24.39	163			65	19.82	222
		162	34.72	162			120	34.90	220
		250	50.45	161			155	43.68	219
		350	65.04	161			190	51.40	217
		441	75.07	161			254	63.92	216
		586	85.88	162			320	74.18	217
			Mean,	164			420	84.87	218
		Av. dev., 2.07%.						Mean,	219
							Av. dev., 0.96%.		
1B	25	15	6.17	290	5B	30	9	4.93	381
		36	14.60	292			29	15.07	374
		63	24.80	289			50	25.22	371
		105	39.70	290			71	35.05	373
		165	57.39	287			107	49.79	370
		235	73.24	289			153	65.34	370
		360	88.84	290			190	75.07	373
			Mean,	290			246	84.93	374
		Av. dev., 0.34%.						Mean,	373
							Av. dev., 0.60%.		
7B	35	8	5.46	482					
		24	15.76	473					
		42	26.59	467					
		60	36.91	467					
		90	52.11	465					
		120	64.63	464					
		155	76.02	467					
		195	84.93	471					
			Mean,	469					
		Av. dev., 0.91%.							

laboratory by Vosburgh and Nelson,[14] it seemed inadvisable to repeat this work. Accordingly the effect of temperature on the shape of the hydrolysis curve was tested by applying Equation 12 to these experiments, with the results shown in Table VI.

In all of these experiments the first sample was taken at a point considerably below 10% inversion. Therefore, in order to compare the average deviation with that which might be due to experimental error, the values obtained from the first sample should be omitted in taking the mean. If this is done the values for the average deviation are as follows: Expt. 11B, 0.68%; 15B, 0.91%; 1B, 0.34%; 5B, 0.43%; and 7B, 0.56%. All of these except that for Expt. 15B are less than 0.7%, and accordingly Equation 12 fits these experiments fairly well. This establishes for the first time the fact that temperature differences, at least between 15° and 35°, have no effect on the shape of the hydrolysis curve or the form of the function, $nt = F(p)$. In other words, an increase in the temperature has the same quantitative effect as an increase in the amount of the invertase used.

VI. Effect of Hydrogen-ion Concentration.

In his classical study of the effect of hydrogen-ion concentration on invertase action, Sörensen[15] found that the velocity coefficient k calculated according to the unimolecular law in the form

$$k = \frac{1}{t_2 - t_1} ln \frac{a - x_1}{a - x_2}$$

increased considerably as the reaction progressed in nearly neutral solutions ($C_{H+} = 10^{-6}$ to 10^{-7}), increased less around the optimum ($C_{H+} = 10^{-4}$ to 10^{-5}), remained constant in slightly more acid solutions ($C_{H+} = 1.2 \times 10^{-4}$), and decreased in still more acid solutions ($C_{H+} = 2.1 \times 10^{-4}$). This means that the shape of the hydrolysis curve or the form of the function $nt = F(p)$ was not the same, in his experiments, for different hydrogen-ion concentrations. Michaelis and Davidsohn[12] have pointed out that this variation may be explained in part by destruction of the invertase in the more acid solutions at the rather high temperature, 52°, at which Sörensen carried on his experiments. By using a lower temperature, 22.3°, they obtained values of k calculated from the equation

$$k = \frac{1}{t} \log \frac{a}{a - x}$$

which increased in experiments at hydrogen-ion concentrations less than 3.0×10^{-3}, where they remained constant. Nelson and Vosburgh,[1] on the other hand, found that at 37° the values of k increased in experiments

[14] Vosburgh and Nelson, "The Temperature Coefficient of Invertase Action," (to be published later).

[15] Sörensen, *Biochem. Z.*, **21**, 131–304 (1909); also *Compt. rend. Lab. Carlsberg*, **8**, 1 (1909).

at the optimum hydrogen-ion concentration, 3.2×10^{-5}, but increased more slowly or remained constant at 3.2×10^{-6}. They noticed, however, that there were some changes in the hydrogen-ion concentration during the latter half of the inversion at about 3.2×10^{-6}.

The equation of the present work will not fit experiments in which the values of the unimolecular "k" are constant or decrease, because it was derived for experiments for which the unimolecular "k" increased. In order to test the effect of hydrogen-ion concentration on the shape of the curve the equation was applied to some recent experiments of Vosburgh and Nelson (to be published later) in which the hydrogen-ion concentration was held constant at 10^{-6} moles per liter by means of citrate buffers and the improved procedure recommended by Vosburgh[16] was used. These results are given in Table VII.

TABLE VII.

EFFECT OF A DIFFERENT HYDROGEN-ION CONCENTRATION. (EXPERIMENTS OF VOSBURGH AND NELSON.)

Initial sucrose concentration, 10 g. per 100 cc. Hydrogen-ion concentration, 1.10×10^{-6} to 1.13×10^{-6} moles per liter. Invertase preparation No. 8, 1 cc. per 100 cc.

Expt.	Temp. °C.	Time, t. Min.	Amt. inverted, p, %.	$n \times 10^5$.	Expt.	Temp. °C.	Time, t. Min.	Amt. inverted, p, %.	$n \times 10^5$.
13B	15	23	4.99	153	2B	25	30	11.10	264
		75	15.43	148			55	19.76	261
		126	25.34	148			80	28.13	261
		183	35.67	148			101	34.84	260
		282	51.57	147			138	45.58	258
		390	65.46	146			217	64.87	258
		490	75.43	146			350	84.69	261
		642	85.52	146				Mean,	260
			Mean,	148				Av. dev., 0.62%.	
			Av. dev., 1.01%.						
9B	35	10	5.88	415					
		26	15.01	415					
		44	24.75	413					
		64	35.10	414					
		98	50.56	412					
		135	64.45	411					
		171	74.96	413					
		225	85.52	416					
			Mean,	414					
			Av. dev., 0.34%.						

Except for the first value in Expt. 13B, the deviation of which may be due to a slight experimental error, as has been already pointed out, the constancy of n is very satisfactory. This means that at a hydrogen-ion concentration of 10^{-6} moles per liter the curve has the same shape or the

[16] Vosburgh, "Some Errors in the Study of Invertase Action," *J. Am. Chem. Soc.*, **43**, 1693 (1921).

function $nt = F(p)$ has the same form as at the optimum hydrogen-ion concentration. The differences found by Nelson and Vosburgh[1] at C_H^+ 3.2×10^{-6} must be ascribed to changes in the hydrogen-ion concentration or in the amount of active invertase due to the use of hydrochloric acid without buffer. The nature of the buffer, however, does not seem to affect the shape of the curve, for the experiments of the present work were made with a 0.01 M buffer mixture of acetic acid and sodium acetate, while the experiments of Vosburgh and Nelson quoted in Tables VI and VII were made with a similar concentration of citric acid and sodium citrate. Very recently the range for which the equation holds has been extended to C_H^+ 3.2×10^{-7} by some experiments of Nelson and Bloom-field (not yet published). These results mean that within the limits given changes in hydrogen-ion concentration affect the activity of the invertase in just the same way as changes in temperature or in the amount of invertase used; either there is actually a change in the amount of the active substance present throughout the experiment or else the activity of the amount present is uniformly reduced or increased by the change and remains constant throughout the experiment.

VII. A Criterion of Normal Invertase Action.

Above, in Part II of this paper, experiments were given which showed that not all invertase preparations impart the same shape to the hydrolysis curve. Equation 12 was made to fit the experiments in Table IA, made with invertase preparations which were classified as normal. Accordingly it seemed probable that it would not fit the experiments in Table IB, and hence might be used as a means of distinguishing between normal and

TABLE VIII.

APPLICATION OF THE EQUATION AS A CRITERION OF NORMAL INVERTASE ACTION.

Expts. B12–15.

Expts. B17 and B18.

1.905 cc. of Invertase 3 per 100 cc.

10.45 cc. of Invertase 6 per 100 cc.

Time $t.$ Min.	Amt. inverted, p, %.	$n\times10^5$.	Time, $t.$ Min.	Rotation, Degrees. B17	B18	Amt. inverted, p, %.	$n\times10^5$.
0	0	...	0	13.15	13.15	0	...
5	3.20	450	15	11.47	11.46	10.03	476
10	6.35	449	30	9.88	9.88	19.41	470
15	9.44	448	50	7.90	7.90	31.16	466
22	13.65	445	70	6.10	6.10	41.84	461
30	18.46	446	90	4.52	4.49	51.34	457
60	34.96	440	115	....	2.75	61.72	454
90	49.38	436	140	1.35	1.32	70.15	451
120	61.48	433	180	−0.36	−0.40	80.30	448
180	78.81	431	240	−1.95	−1.96	89.67	449
300	93.59	426	300	−2.74	−2.74	94.30	443
	Mean, 440		2 to 7 days	−3.71	−3.71	Mean,	457
	Av. dev., 1.4%.					Av. dev., 1.9%.	

abnormal invertase preparations. Equation 12, therefore, was applied to the results of Table IB, and also to experiments with two other invertase preparations, Nos. 6 and 7, with the results shown in Table VIII.

Expts. B20 and B21.
3.60 cc. of Invertase 7 per 100 cc.

Time, $t.$ Min.	B20.	Rotation, Degrees. B21.	Amt. inverted, $p, \%$.	$n \times 10^5$.
0	13.07	13.08	0	...
15	11.47	11.47	9.55	453
30	9.89	9.91	18.87	456
50	7.94	7.96	30.45	454
70	6.17	6.16	41.00	451
90	4.53	4.54	50.68	450
117	2.64	2.66	61.90	448
140	1.29	1.33	69.85	448
180	−0.49	−0.43	80.36	449
240	−2.06	−2.06	89.85	452
300	−2.85	−2.81	94.42	446
2 to 7 days	−3.78	−3.78		Mean, 450

Av. dev., 0.56%.

It will be noticed in Expts. B12–15, Table VIII, that for those points where the data of Table IB coincided with those of Table IA, or up to 20% of the inversion, the values of n are fairly constant, even for abnormal invertase. The abnormality, however, shows up later in the decreasing values of n, and is indicated by the larger values of the average deviation of a single value of n from the mean, which is well above 0.7% for the experiments with Invertase 3. Expts. B17 and B18 indicate that Invertase 6 is also an abnormal invertase preparation, since the values of n decrease and the average deviation is well above 0.7%. Invertase 7, on the other hand, is a normal invertase preparation, as is shown by Expts. B20 and B21, since the values of n exhibit satisfactory constancy. These experiments indicate that Equation 12 may be used as a criterion of normal invertase action. In order to decide whether invertase preparations are normal or abnormal, then, it is no longer necessary to use them at initially equivalent effective concentrations, but the experiments may be made with any concentration, at least within the limits of the experiments in Table V. If the average deviation of the values of n is under 0.7%, the invertase preparation may be classified as normal; if the values of n decrease and the average deviation is much over 0.7%, then the invertase preparation is abnormal.

VIII. Attempts to Make the Abnormal Invertase Act Normally.

There were no known differences in the method used in obtaining the normal and abnormal invertase preparations. However, it was deemed advisable to find out whether the abnormality could be due to some impurity which might be removed by further dialysis. Accordingly a sample of Invertase 6 was dialyzed for 3 days more in a collodion bag against running tap water. During the dialysis its volume was about doubled and its activity decreased by about $1/2$ on that account; this was designated as Invertase 6B. To avoid this loss in activity, a sample of Invertase 3 was concentrated by evaporation in a collodion bag by fanning at room temperature[17] until it had lost about half its volume, and then dialyzed for 4 days, when it had regained about its original volume; this was designated as Invertase 3B. The results of experiments with these dialyzed preparations are given in Table IX.

TABLE IX.

EFFECT OF DIALYSIS ON ABNORMAL INVERTASE.

Expt. B22.				Expt. B24.			
16 cc. of Invertase 6B per 100 cc.				1.903 cc. of Invertase 3B per 100 cc.			
Time, t. Min.	Rotation, degrees.	Amt. inverted, p, %.	$n \times 10^5$.	Time, t. Min.	Rotation, degrees.	Amt. inverted. p, %.	$n \times 10^5$.
0	13.10	0	...	0	13.05	0	...
15	11.41	10.03	476	15	11.44	9.55	453
30	9.78	19.70	477	30	9.89	18.75	453
50	7.74	31.81	476	60	7.08	35.43	446
70	5.89	42.79	473	90	4.63	49.97	442
90	4.24	52.58	470	120	2.65	61.72	435
115	2.53	62.73	464	180	−0.19	78.58	429
140	1.10	71.22	462	240	−1.77	87.95	422
180	−0.58	81.19	459	300	−2.61	92.94	412
240	−2.08	90.09	456	4 days	−3.80	Mean,	437
300	−2.83	94.54	449			Av: dev., 2.75%.	
		Mean,	466				
		Av. dev., 1.76%.					

The decrease in the values of n and the large average deviations show that the invertase was still abnormal.

Since the abnormality could not be removed by purification by dialysis, it was thought that it might be due to the absence of some substance contained in the normal invertase. A sample of Invertase 8 was inactivated by boiling, and was proved to be totally inactive by the absence of any action on sugar solutions. Experiments were then conducted in which the solutions contained 10 cc. of this inactive invertase per 100 cc. in addition to the abnormal invertase under investigation. The results are given in Table X.

The figures in Table X indicate that the presence of boiled normal invertase caused preparation No. 3 to act normally, giving constant values of n, while it was practically without effect on preparation No. 6. · This apparently means that there are different kinds of abnormality in different invertase preparations.

[17] Kober, J. Am. Chem. Soc., **39**, 944 (1917).

TABLE X.

EFFECT OF BOILED NORMAL INVERTASE ON THE ACTION OF ABNORMAL INVERTASE.

Expt. B25.		Expt. B26.	
10 cc. of boiled Invertase 8 and 1.905 cc. of Invertase 3 per 100 cc.		10 cc. of boiled Invertase 8 and 1.943 cc. of Invertase 3 per 100 cc.	

Time, t. Min.	Rotation, degrees.	Amt. inverted, p, %.	$n \times 10^5$.	Time, t. Min.	Rotation, degrees.	Amt. inverted, p, %.	$n \times 10^5$.
0	13.17	0	...	0	13.17	0	...
15	11.61	9.26	439	15	11.59	9.38	445
30	10.10	18.22	440	30	10.06	18.46	446
60	7.31	34.78	437	60	7.24	35.19	443
90	4.85	49.38	436	90	4.76	49.91	441
120	2.78	61.66	434	120	2.67	62.31	441
180	−0.19	79.29	437	180	−0.26	79.70	441
240	−1.85	89.14	440	240	−1.91	89.50	446
300	−2.68	94.07	437	300	−2.72	94.30	443
3 days	−3.68	Mean, 437		3 days	−3.71	Mean, 443	
		Av. dev., 0.34%.				Av. dev., 0.40%.	

Expts. B36 and B38.		Expts. B35 and B37.	
10.45 cc. of Invertase 6 per 100 cc.		10 cc. of boiled Invertase 8 and 10.45 cc. of Invertase 6 per 100 cc.	

Time, t. Min.	Rotation, Degrees. B36.	Rotation, Degrees. B38.	Amt. inverted, p, %.	$n \times 10^5$.	Time, t. Min.	Rotation, Degrees. B35.	Rotation, Degrees. B37.	Amt. inverted, p, %.	$n \times 10^5$.
0	13.14	13.13	0	...	0	13.25	13.24	.....	...
15	11.55	11.57	9.38	445	15	11.67	11.67	9.38	445
30	10.04	10.06	18.34	443	30	10.15	10.16	18.34	443
60	7.27	7.29	34.78	437	60	7.36	7.39	34.84	438
90	4.87	4.88	49.02	432	90	4.94	4.98	49.20	434
120	2.84	2.84	61.13	429	120	2.92	2.94	61.25	430
180	−0.04	−0.06	78.28	426	180	−0.01	0.04	78.52	428
240	−1.72	−1.71	88.13	424	240	−1.67	−1.64	88.43	429
300	−2.61	−2.59	93.41	422	300	−2.53	−2.50	93.53	425
3-6 days.	−3.72	−3.72	Mean, 433		2-5 das.	−3.60	−3.61	Mean, 434	
			Av. dev., 1.67%.					Av. dev. 1.38%.	

In order to show that preparation No. 3 had not become normal simply on standing, but that the normal course of the reaction was really produced by the presence of the boiled invertase, other experiments were run with Invertase 3 alone, and were found to give decreasing values of n, as before. These values, however, were somewhat smaller than those obtained in Expts. B12 to B15, indicating that this preparation had appreciably lost activity on being kept in the ice-box for less than 5 months.

Further experiments were made with the abnormal invertase preparations Nos. 3 and 6 in the presence of different concentrations of sodium chloride. The results are given in Table XI.

The figures in Table XI indicate that increasing concentrations of sodium chloride exert an increasing retarding effect on the action of the

TABLE XI.

EFFECT OF SODIUM CHLORIDE ON ABNORMAL INVERTASE.

Expt. B44.
1.905 cc. of Invertase 3 per 100 cc. in 0.02 M NaCl.

Expt. B43.
1.905 cc. of Invertase 3 per 100 cc. in 0.05 M NaCl.

Time, t. Min.	Rotation, degrees.	Amt. inverted, p, %.	$n \times 10^5$.	Time, t. Min.	Rotation, degrees.	Amt. inverted, p, %.	$n \times 10^5$.
0	13.05	.....	...	0	13.05	0	...
17	11.34	10.15	425	15	11.54	8.96	424
30	10.08	17.63	425	30	10.12	17.39	419
60	7.38	33.65	422	60	7.44	33.29	417
90	4.99	47.83	420	90	5.08	47.30	414
120	2.99	59.70	416	120	3.03	59.47	414
180	0.00	77.45	417	180	0.05	77.15	414
240	−1.73	87.72	419	240	−1.69	87.48	415
300	−2.64	93.14	416	300	−2.61	92.94	412
2–3 days	−3.80	Mean, 420		2–4 days	−3.82	Mean, 416	
		Av. dev., 0.71%.				Av. dev., 0.70%.	

Expt. B42.
1.905 cc. of Invertase 3 per 100 cc. in 0.1 M NaCl.

Expt. B48.
10.45 cc. of Invertase 6 per 100 cc. in 0.1 M NaCl.

Time, t. Min.	Rotation, degrees.	Amt. inverted, p, %.	$n \times 10^5$.	Time, t. Min.	Rotation, degrees.	Amt. inverted, p, %.	$n \times 10^5$.
0	13.05	0	...	0	13.13	0	...
15	11.58	8.72	413	15	11.61	9.02	427
30	10.16	17.15	413	30	10.17	17.57	423
60	7.50	32.94	410	60	7.52	33.29	417
90	5.14	46.94	410	90	5.19	47.12	412
120	3.13	58.87	408	123	3.02	60.00	408
180	0.14	76.62	408	165	0.86	72.82	407
240	−1.66	87.30	413	225	−1.17	84.87	408
300	−2.59	92.82	410	300	−2.47	92.58	405
3 days	−3.84	Mean, 411		3–6 days	−3.75	Mean, 414	
		Av. dev., 0.46%.				Av. dev., 1.67%.	

invertase. This was not observed in the work of Fales and Nelson[18] at the optimum hydrogen-ion concentration, but this may be due to the fact that they worked with a very much smaller sugar concentration, 0.5 g. per 100 cc. This retardation, however, seems to have more effect at the beginning of the hydrolysis than at the end in the case of Invertase 3, for in Expt. B42, with 0.1 M sodium chloride, the values of n were constant and the action must be classed as that of normal invertase. Invertase 6, however, was not made normal by 0.1 M sodium chloride, for in Expt. B48 the values of n decreased as much as ever. An experiment with 0.5 M sodium chloride and Invertase 6 gave values which decreased somewhat less, but still were not constant enough for the action to be regarded

[18] Fales and Nelson, *J. Am. Chem. Soc.*, **37**, 2769 (1915).

as normal. Unfortunately the supply of Invertase 6 became too low for further experiments to be carried out with it.

Further experiments were made with Invertase 3 to determine the effect of invertase concentration on the abnormal action. The results are given in Table XII.

TABLE XII.

ABNORMAL INVERTASE AT DIFFERENT CONCENTRATIONS.

| Expts. B58 and B59. | | | | | Expts. B54 and B55. | | | | |
| 0.5 cc. of Invertase 3 per 100 cc. | | | | | 3 cc. of Invertase 3 per 100 cc. | | | | |
Time, t. Min.	Rotation, Degrees. B58.	B59.	Amt. inverted, p, %.	$n \times 10^4$.	Time, t. Min.	Rotation, Degrees. B54.	B55.	Amt. inverted, p, %.	$n \times 10^4$.
0	13.05	13.05	0	...	0	13.06	13.06	0	...
60	11.53	11.51	9.08	108	10	11.46	11.48	9.44	671
120	10.06	10.05	17.80	107	20	9.96	9.98	18.34	664
195	8.35	8.33	27.95	106	30	8.52	8.55	26.88	662
270	6.72	6.77	37.45	106	45	6.50	6.53	38.87	660
360	5.06	5.06	47.42	104	70	3.62	3.67	55.91	654
450	3.50	3.54	56.56	103	100	0.98	....	71.69	654
540	2.22	2.20	64.33	102	120	−0.29	−0.28	79.23	654
1101	−1.96	....	89.08	96	150	−1.62	−1.62	87.12	656
7 to 12 days			Mean, 104		3 to 8 days			Mean, 659	
	−3.79	−3.79	Av. dev., 2.7%			−3.79	−3.79	Av. dev., 0.74%	

Expts. B56 and B57.

6 cc. of Invertase 3 per 100 cc.

Time, t. Min.	Rotation, Degrees. B56.	B57.	Amt. inverted, p, %.	$n \times 10^4$.
0	13.12	13.12	0	...
6	11.21	11.23	11.28	134
12	9.40	9.43	22.02	134
18	7.72	7.72	32.05	133
26	5.66	5.66	44.27	133
35	3.62	3.62	56.38	132
48	1.25	1.30	70.33	132
65	−0.88	−0.84	82.97	134
85	−2.26	......	91.28	135
4 to 5 days	−3.75	−3.74	..... Mean, 133	
			Av. dev., 0.67%.	

These results show that in the case of Invertase 3 the abnormality decreases with increasing amount of invertase or increases with decreasing amount of invertase or increasing time of reaction.

It is not possible to explain these changes in the abnormality of Invertase 3 at the present time.

IX. Attempts to Make Normal Invertase Become Abnormal.

Since the presence of sodium chloride had seemed to some extent to favor the normal course of invertase action, it seemed worth while to find out whether further dialysis, by removing any last traces of salt, could make a specimen of normal invertase act abnormally. Accordingly a sample of Invertase 8 was dialyzed for a week in a collodion bag against 8 changes of distilled water. This was designated as Invertase 8A, and when tested was found to be still normal, as is shown by the results in Table XIII, Expt. B46.

Another sample of Invertase 8 was partially inactivated by heating for 1 hour on a water-bath at 50°, and then for $1/2$ hour more at about 57°. This reduced its activity by about one-half. This invertase, No. 8E, was also found to be still normal, as is shown by Expts. B50 and B51, Table XII.

A further attempt to render Invertase 8 abnormal was made by exposing some of it for 2 hours, in a quartz flask, to the ultraviolet and other radiation given by a mercury arc lamp. The result, Invertase 8F, had about one-half the activity of Invertase 8, but was also found to be still normal, as is shown by Expts. B52 and B53, Table XIII.

TABLE XIII.

ACTION OF NORMAL INVERTASE AFTER FURTHER DIALYSIS, HEATING, AND EXPOSURE TO THE MERCURY ARC.

Expt. B46. Invertase 8A, dialysed, 5 cc. per 100 cc.				Expts. B50 and B51. Invertase 8E, heated, 5 cc. per 100 cc.			
Time, t, Min.	Rotation, Degrees.	Amt. inverted, p, %.	$n \times 10^5$.	Time, t, Min.	Rotation, B50.	B51. Degrees.	Amt. inverted p, %.	$n \times 10^5$.
0	13.06	0	...	0	13.10	13.10	0	...
15	11.27	10.62	505	15	10.94	10.94	12.82	612
30	9.56	20.77	504	30	8.88	8.89	25.04	614
60	6.44	39.29	501	50	6.37	6.38	39.94	612
90	3.78	55.07	499	70	4.18	4.19	52.94	610
120	1.66	67.66	497	90	2.30	2.35	63.98	610
180	−1.14	84.27	501	110	0.79	0.84	72.94	612
230	−2.34	91.39	502	140	−0.87	−0.85	82.85	618
270	−2.87	94.54	499	180	−2.20	−2.18	90.74	624
3 days	−3.79	Mean,	501	2-5 days	−3.78	−3.78	Mean,	614
	Av. dev., 0.40%.					Av. dev., 0.57%.		

Expts. B52 and B53.

Invertase 8F, exposed to mercury arc, 5 cc. per 100 cc.

Time, t, Min.	B52.	Rotation, B53. Degrees.	Amt. inverted, p, %.	$n \times 10^5$.
0	13.09	13.09	0	...
15	10.33	10.35	16.32	785
25	8.61	8.63	26.53	783
35	6.99	7.04	36.08	781
45	5.51	5.53	44.92	779
65	2.93	2.99	60.12	775
85	0.94	0.97	72.05	776
105	−0.54	−0.53	80.89	781
140	−2.12	−2.12	90.27	787
1 to 3 days	−3.77	−3.80	.....	Mean, 781
				Av. dev., 0.38%.

Since the values of n in Table XIII are constant in each experiment, having an average deviation from the mean in each case of less than 0.7%, the results show that the invertase was still acting normally. Hence it may be concluded that it is not possible by any of these three methods of treatment to render a normal invertase preparation abnormal.

X. Experimental Details.

Preparation of materials.—The invertase used was all obtained from yeast by the method of Nelson and Born,[19] with slight modifications as described below. Preparations 6 and 7 had been made by previous workers in this laboratory and had been kept for several years in solution, saturated with toluene, in an ice-box. Preparations 1, 2 and 3 were made from yeast which had been permitted to autolyze for about a month and then filtered, and the filtrate had been treated with toluene and kept in stoppered bottles at room temperature for three years or more. During this time more solid matter had separated, and this was filtered off and the filtrate treated according to the method described by Nelson and Born[19] with the following modifications. Only one precipitation with alcohol was used and the kaolin treatment was omitted. After treatment with lead acetate and potassium oxalate, the filtrate was dialyzed for from 4 to 6 days in collodion bags against running tap water. The solutions become colorless and nearly clear during the dialysis. The dialyzed solutions were not precipitated again, but were preserved with toluene and kept in the ice-box until needed for the experiments. Preparation No. 8 was prepared by the same method from a new lot of 100 pounds of pressed yeast.[20] The preparation of Invertase No. 2 was carried out by Nelson and Simons,[21] who modified the treatment further by nearly neutralizing the solution with ammonia before the alcohol precipitation. No differences in the method of preparation are known which might account for the abnormality of invertase preparations Nos. 3 and 6.

Two lots of sucrose were used. In each case the starting point was the best commercial sugar, which was dissolved in distilled water and clarified with charcoal. The first lot was precipitated by alcohol by the method of Cohen and Commelin.[22] Its rotation was found to agree within 0.1% with that calculated from the formulas of Landolt and Schönrock.[23] The second lot was recrystallized from water by a procedure similar to that of Bates and Jackson.[24] Its rotation agreed with the calculated value within 0.04%.

Other chemicals were c. p. grades, used without further purification.

[19] Nelson and Born, *J. Am. Chem. Soc.*, **36**, 393 (1914).

[20] Kindly furnished by the Jacob Ruppert Brewery of New York City.

[21] Simons, *Dissertation*, Columbia University, 1921; Nelson and Simons, *J. Am. Chem. Soc.*, (to be published later).

[22] Cohen and Commelin, *Z. physik. Chem.*, **64**, 29 (1908).

[23] Browne, "A Handbook of Sugar Analysis," John Wiley and Sons, New York, 1912, pp. 177-8.

[24] Bates and Jackson, *Bur. of Standards Sci. Papers*, No. **268**, 75 (1916).

Apparatus.—Constant temperature was obtained by the use of an electrically controlled water-bath which remained at $25° \pm 0.01°$.

The progress of the inversion was followed by means of a Schmidt and Haensch polarimeter reading to $0.01°$. The tubes used were 200 mm. long, and were proved to be of the same length by observing the rotation of the same 10% sugar solution in each tube. The temperature of the tubes was kept constant by the thermostat described by Nelson and Beegle,[25] which maintained a temperature of $25° \pm 0.05°$.

Monochromatic light of wave length 546.1 $\mu\mu$ was obtained from a mercury vapor arc by purification through two Wratten filters, one a No. 77, and the other a No. 77 which had been re-cemented with a green film in place of the yellow one. Thanks are due to Dr. C. E. K. Mees of the Eastman Kodak Company for preparing these filters. This light made it possible to use the polariscope with a half-shadow angle of $0.5°$.

Nonsol bottles were used to contain the solutions undergoing hydrolysis. All volumetric apparatus used in making up solutions was calibrated.

Control of the Hydrogen-ion Concentration.—The desired hydrogen-ion concentration was obtained by the use of a buffer mixture of 0.1 M acetic acid and 0.1 M sodium acetate in the proportions given by Michaelis.[26] One hundred cc. of the final solution always contained 10 cc. of this buffer, making the total concentration $0.01 \, M$. This concentration was low enough so that any salt effect on the invertase action was negligible, especially at the optimum hydrogen-ion concentration.[18] In the experiments of Nelson and Vosburgh[1] the desired hydrogen-ion concentration was obtained by the use of diluted hydrochloric acid. In the experiments of Vosburgh and Nelson[14] a buffer of citric acid and secondary sodium citrate was used at a total citrate concentration of 0.01 M.

The hydrogen-ion concentration was measured during or after each inversion by the colorimetric method of Sörensen,[15] using α-naphthyl-amino-azo-p-benzene sulfonic acid as indicator with citrate standards. The latter were standardized electrometrically with the hydrogen electrode and the saturated potassium chloride calomel cell[27] using a salt bridge of saturated potassium chloride solution. The hydrogen-ion concentrations were based on 0.1000 M hydrochloric acid as a standard, its ionization[28] being taken as 92.04% at 25,° the temperature at which the present determinations were made.

Procedure.—In general the procedure followed was that recommended by Vosburgh.[16] Duplicate experiments were run on different days with freshly prepared sugar solutions. A solution was made up containing

[25] Nelson and Beegle, *J. Am. Chem. Soc.*, **41**, 559 (1919).
[26] Michaelis, "Die Wasserstoffionenkonzentration," Springer, Berlin, **1914**, p. 184.
[27] Fales and Mudge, *J. Am. Chem. Soc.*, **42**, 2434 (1920).
[28] Fales and Vosburgh, *ibid.*, **40**, 1295 (1918).

sucrose and buffer in such concentrations that when a certain volume of this had been measured out it would be possible to add from a pipet a round number of cubic centimeters of invertase to start the reaction. For example, in Expt. B1 32.680 g. of sucrose and 32.68 cc. of buffer were diluted to 500 cc. at 25°. Of this solution 321.80 cc. was pipetted into a Nonsol bottle, and 5 cc. of invertase was added to start the reaction. This produced the initial concentrations given in Table I A. The solutions were stirred by a current of filtered air while being mixed, and samples were taken by pipets delivering in 10 seconds or less.[21] The time of mixing or of sampling was taken as the mean time of delivery of the pipet used. The reaction was stopped and mutarotation hastened by the use of sodium carbonate as recommended by Hudson,[29] a 25cc. sample being added to 5 cc. of 0.1 M sodium carbonate solution. The initial rotation of each solution was determined by preparing samples of identical composition in which the sodium carbonate was added to the sugar before the addition of the invertase, thus rendering the invertase entirely inactive. The rotation of each solution was determined by taking the mean of at least four concordant readings, the tube being rotated slightly after each reading to ensure the detection of any strain in the cover glasses.[30] The zero point of the polariscope was similarly determined by the use of a tube filled with distilled water. The final rotations were obtained by taking samples 2 days or more after the start of the reaction. Samples taken on the second and third days usually had the same rotation. In calculating the percentage inverted, the total change in rotation was always taken as 16.85°, since this value was obtained in all the experiments of Vosburgh and Nelson[14] as well as in the majority of the present experiments.

Since in several experiments the total change in rotation appeared to be a few hundredths of a degree more than 16.85°, it was thought advisable to test the effect of such differences on the values of n as obtained by the use of Equation 12. A sample calculation was made for Expt. B46, Table XIII, with the following results.

Using 16.85°, $n \times 10^{8}$.		Using 16.89°, $n \times 10^{8}$.	
505	501	504	498
504	502	503	497
501	499	500	492
499	Mean, 501	497	Mean, 498
497	Av. dev., 0.40%	495	Av. dev., 0.60%.

These results show that an error of 0.04° in determining the total change in rotation could not have caused sufficient error in the values of n to make a normal invertase preparation appear abnormal. Since this was the

[29] Hudson, *J. Am. Chem. Soc.*, **30**, 1564 (1908).
[30] Browne, Ref. 23, p. 156.

extreme deviation noticed from the value 16.85°, the procedure adopted of taking the total change as 16.85° in all calculations is quite justified.

A calculation of the possible error in determining the rotation of any sample which might be due to errors in the various measurements of weight and volume involved in this procedure has been made by Messrs. G. Bloomfield and F. Hollander of this laboratory. Using estimates of these errors based on the present authors' calibrations, this calculation gave a maximum error of about 0.01° in the determination of the rotation of a sample. Since the duplicate experiments did not always agree so well as this, a fairer estimate of the precision of the measurements may be obtained from the agreement of the duplicates themselves. This would put the average difference between duplicate measurements of a change in rotation at about 0.02°. The effect of such an error on the values of n obtained by the use of Equation 12 has already been considered.

Summary.

1. It has been shown that not all preparations of yeast invertase are alike in their action, but that some are abnormal in allowing the hydrolysis of cane sugar to slow up more than others after the first 20% of the inversion.

2. An empirical equation is given which fits the hydrolysis of cane sugar by normal invertase over an extreme range of invertase concentration of 12:1. By this means it has been shown that the hydrolysis-time curves for normal invertase are of the same shape for these different invertase concentrations and can be made to superimpose if the time scale be multiplied by the proper constant.

3. By the same method it has been shown that the hydrolysis curve with normal invertase has the same shape at temperatures varying from 15° to 35°, and at hydrogen-ion concentrations from 4.0×10^{-5} to 3.2×10^{-7}.

4. It was found that one abnormal invertase preparation could be rendered normal by the presence of boiled normal invertase or 0.1 M sodium chloride, while another was not affected by either. The former preparation also worked normally at a very high concentration.

5. It was found impossible to render a normal invertase preparation abnormal by further dialysis or partial inactivation by heating or ultraviolet light.

VITA.

David Ingersoll Hitchcock was born in Detroit, Michigan, on June 26, 1893. In 1911 he was graduated from the Detroit Central High School. In 1915 he was graduated from Dartmouth College with the degree of Bachelor of Arts. From 1915 to 1917 he was Instructor in Chemistry at Dartmouth College. He was a graduate student in Chemistry at Columbia University during the summers of 1915, 1916, and 1917. In August, 1917, he enlisted in the 101st Machine Gun Battalion of the 26th Division, United States Army. In June, 1918, he was transferred to the Gas Service, later the Chemical Warfare Service, and was assigned to the chemical laboratory at Hanlon Field, Chaumont, France. In January, 1919, he was discharged from the Army. Since February, 1919, he has continued his studies at Columbia University, where he received the degree of Master of Arts in 1919. He has been laboratory assistant in various courses in the Department of Chemistry. Since February, 1920, he has been Harriman Research Assistant in the laboratory of Professor J. M. Nelson. He is co-author with Professor Nelson of a paper on "The Activity of Adsorbed Invertase," which has been accepted for publication in the *Journal of the American Chemical Society*. For the year 1921–1922 he has been appointed a Fellow of the Rockefeller Institute for Medical Research, New York City.

**THIS BOOK IS DUE ON THE LAST DATE
STAMPED BELOW**

AN INITIAL FINE OF 25 CENTS

WILL BE ASSESSED FOR FAILURE TO RETURN
THIS BOOK ON THE DATE DUE. THE PENALTY
WILL INCREASE TO 50 CENTS ON THE FOURTH
DAY AND TO $1.00 ON THE SEVENTH DAY
OVERDUE.

NOV 2 1939

NOV 2